Let's Find Out About SUBTRACTION

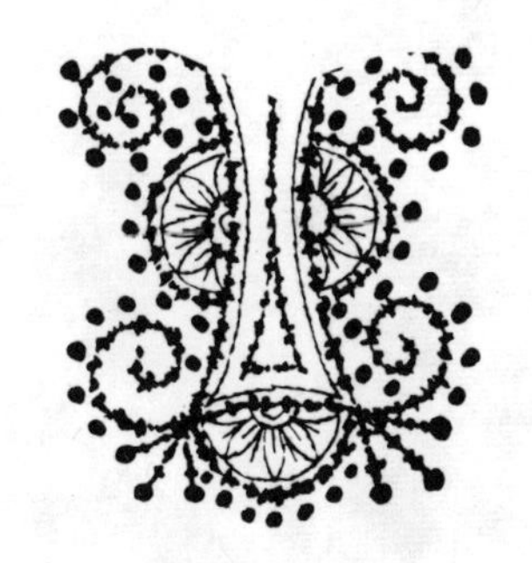

Let's Find Out About

SUBTRACTION

by David C. Whitney

pictures by Eva Cellini

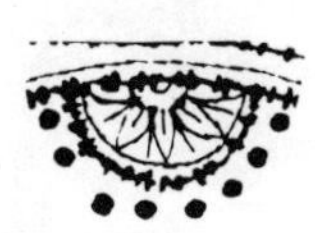

Franklin Watts, Inc.
730 Fifth Avenue New York, N.Y. 10019

SBN 531–00045–1

Library of Congress Catalog Card Number: 68-11890
Printed in the United States of America
by The Moffa Press, Inc.

Let's Find Out About SUBTRACTION

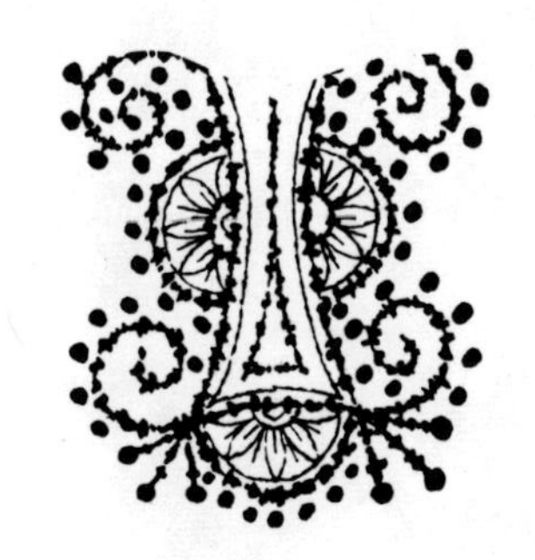

Black is the opposite of white.
Wet is the opposite of dry.

A wet black dog is the opposite of a dry white cat.
And subtraction is the opposite of addition.

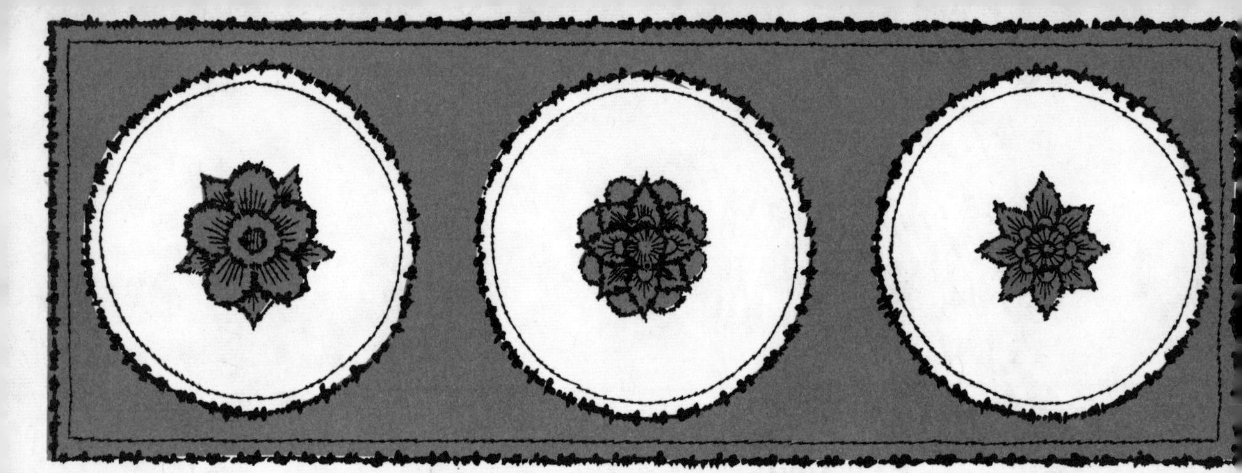

In arithmetic a group of things that are alike in some way is called a set. Each object in the set is called a member.

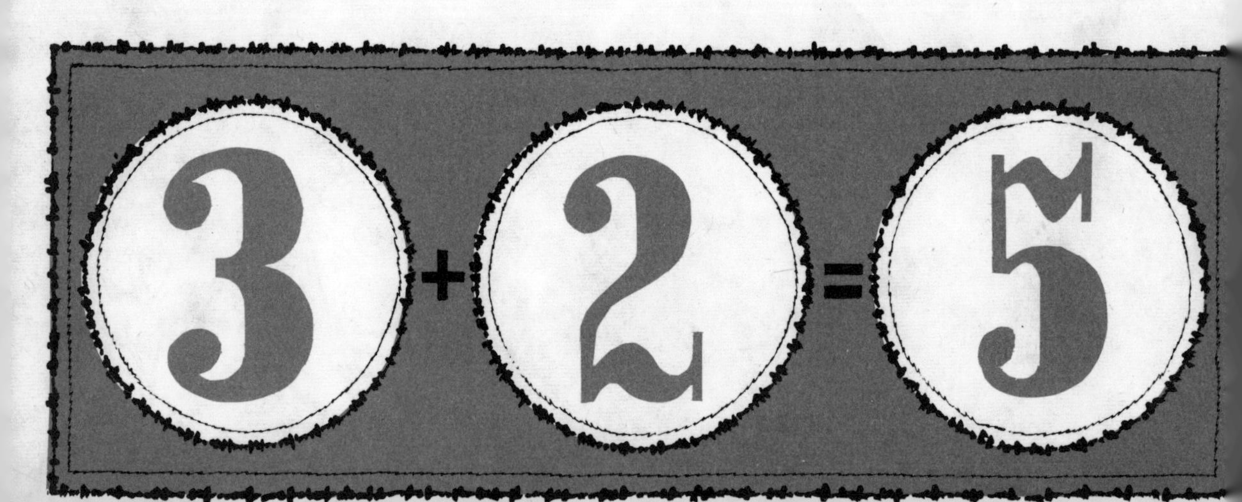

When we use addition, we add the members of sets together to make a greater set.

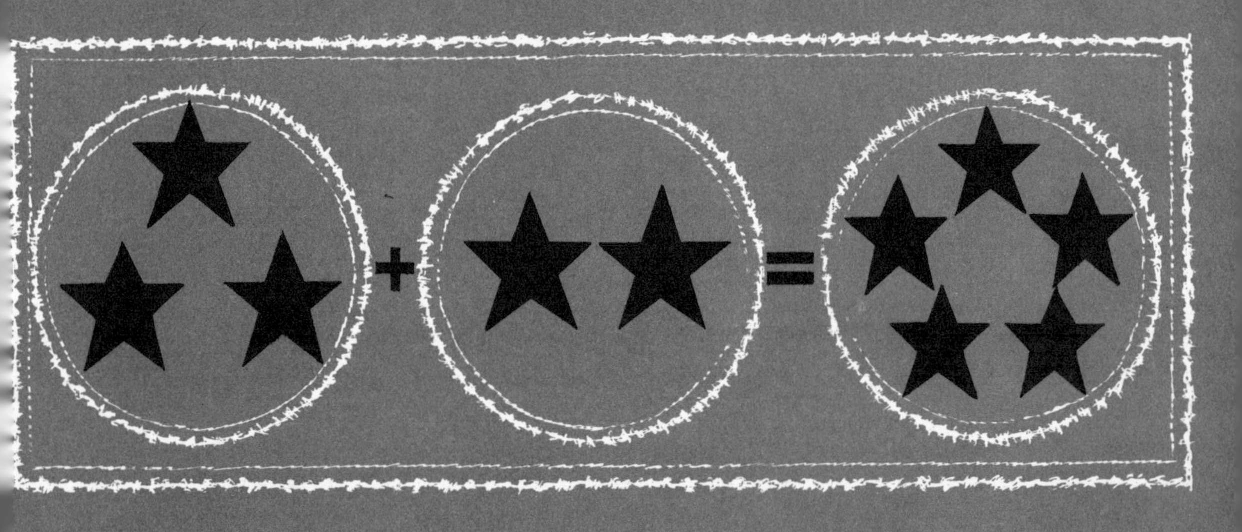

Here we add a set of stars with 3 members to a set of stars with 2 members, and they equal, or add up to, a set of stars with 5 members.

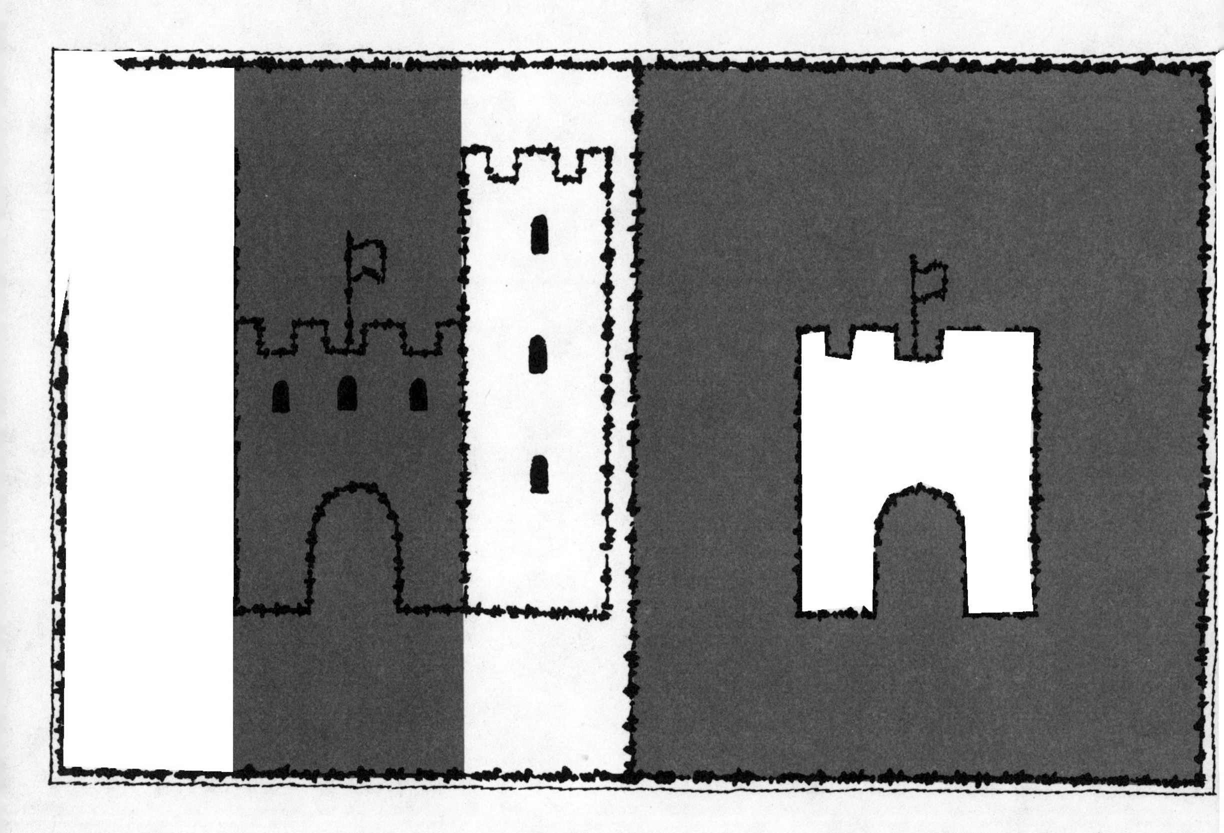

When we use subtraction, we subtract, or take away, the members of one set from the members of another set to make a smaller set.

Here we subtract a set of stars with 2 members from a set of stars with 3 members, and we are left with a set of stars with 1 member.

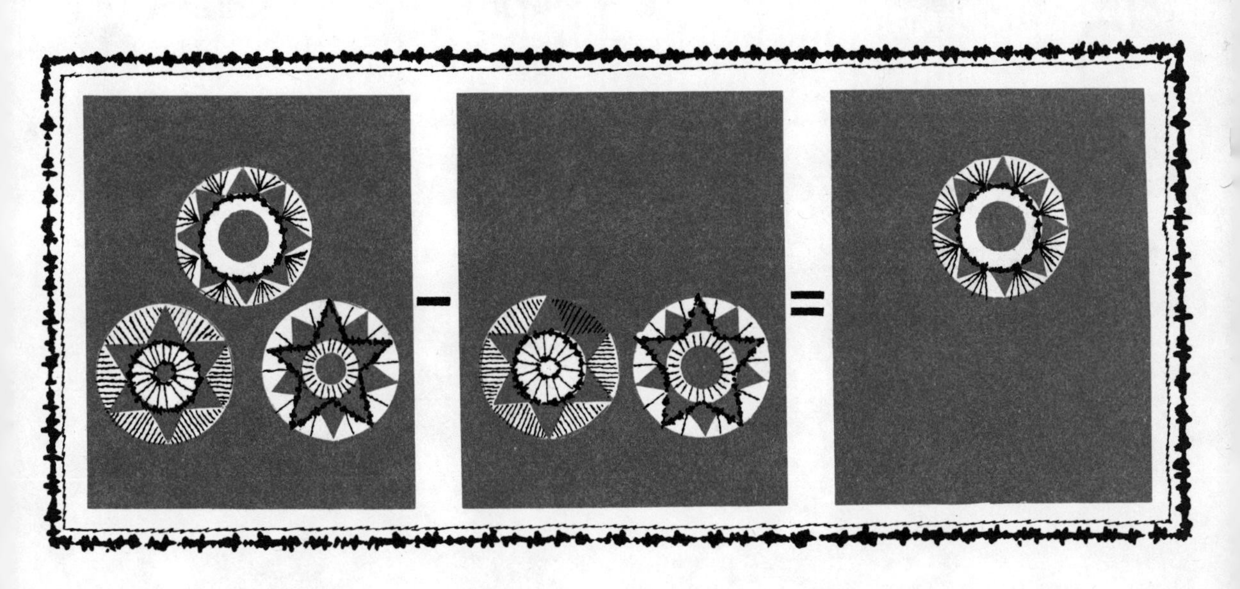

Instead of using stars, we can write this in another way. We can write

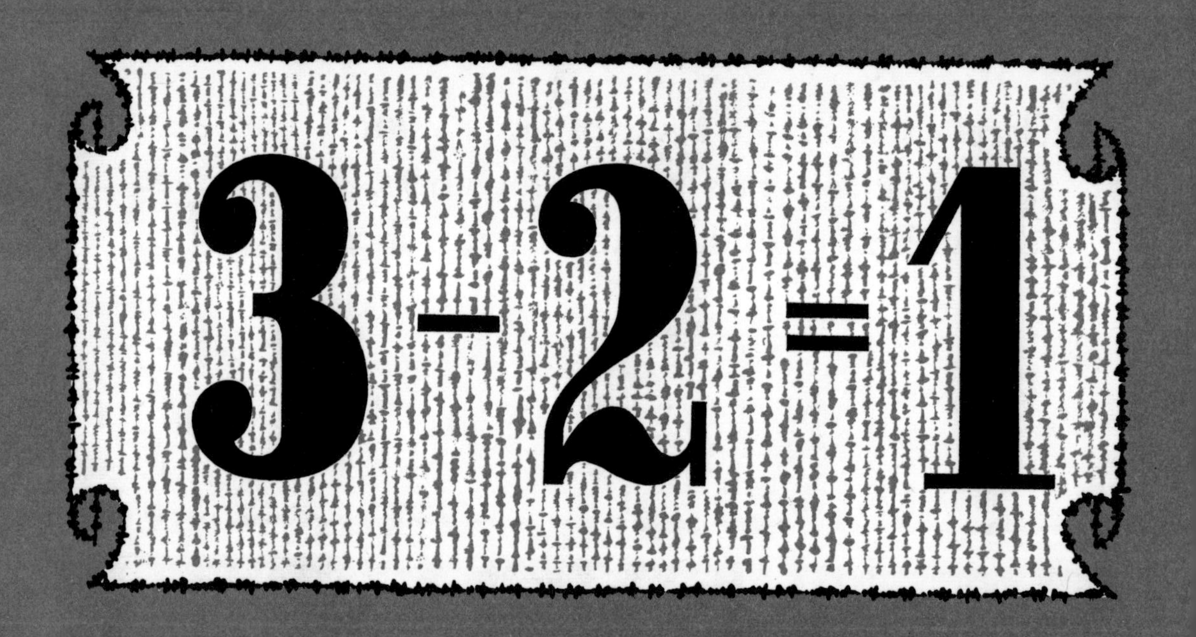

We say it like this — three minus two is one.

Sometimes we write

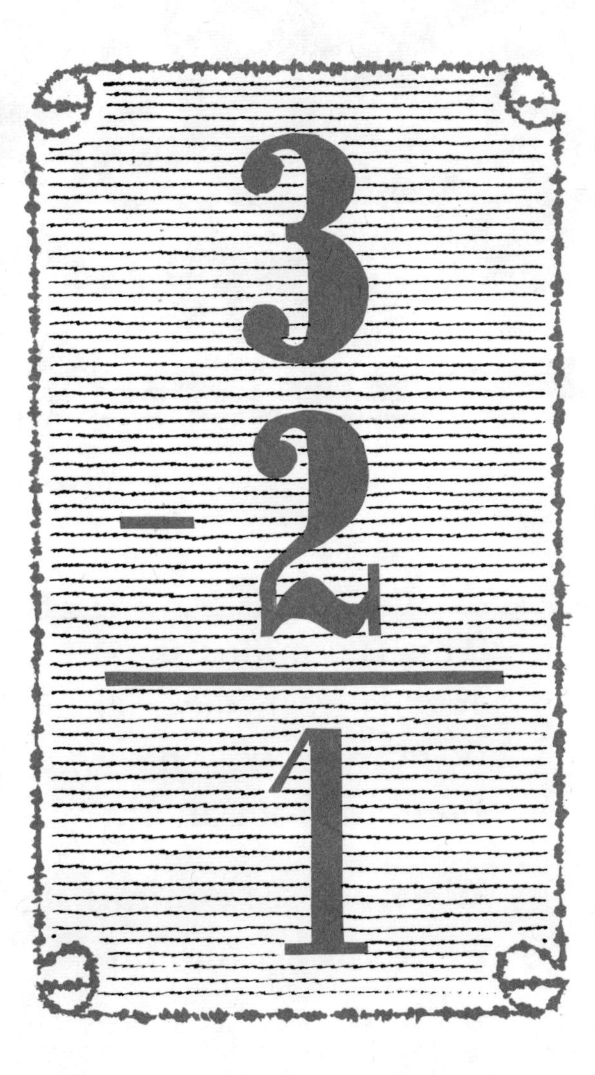

That also says three minus two is one.

3 - 2

This is a minus sign.
It tells us to subtract,
or take away, one
number from another.

When we see a minus sign
we know we should
subtract.

3 - 2 = 1

This is an equal sign.

It shows us what we have left when we subtract one number from another number.

Everyone uses subtraction.

SERVICE

A gas-station man uses subtraction to find out how many gallons of gasoline he has left after he has been filling up cars all day. He subtracts the number of gallons he has used from the number of gallons he had that morning. Then he knows how many gallons he has left.

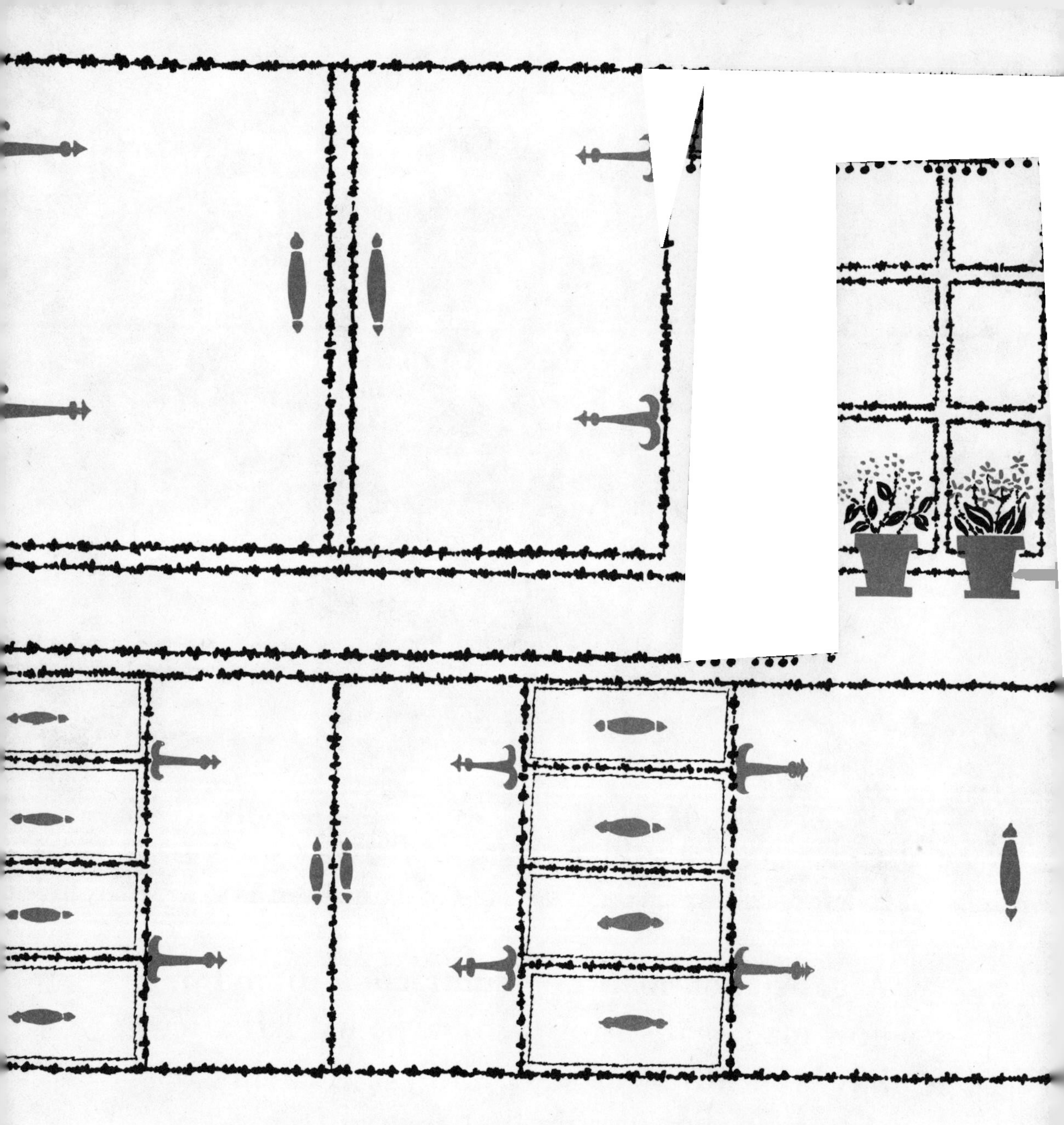

A mother uses subtraction to find out how many eggs she will have left if she uses 4 eggs to make a cake.

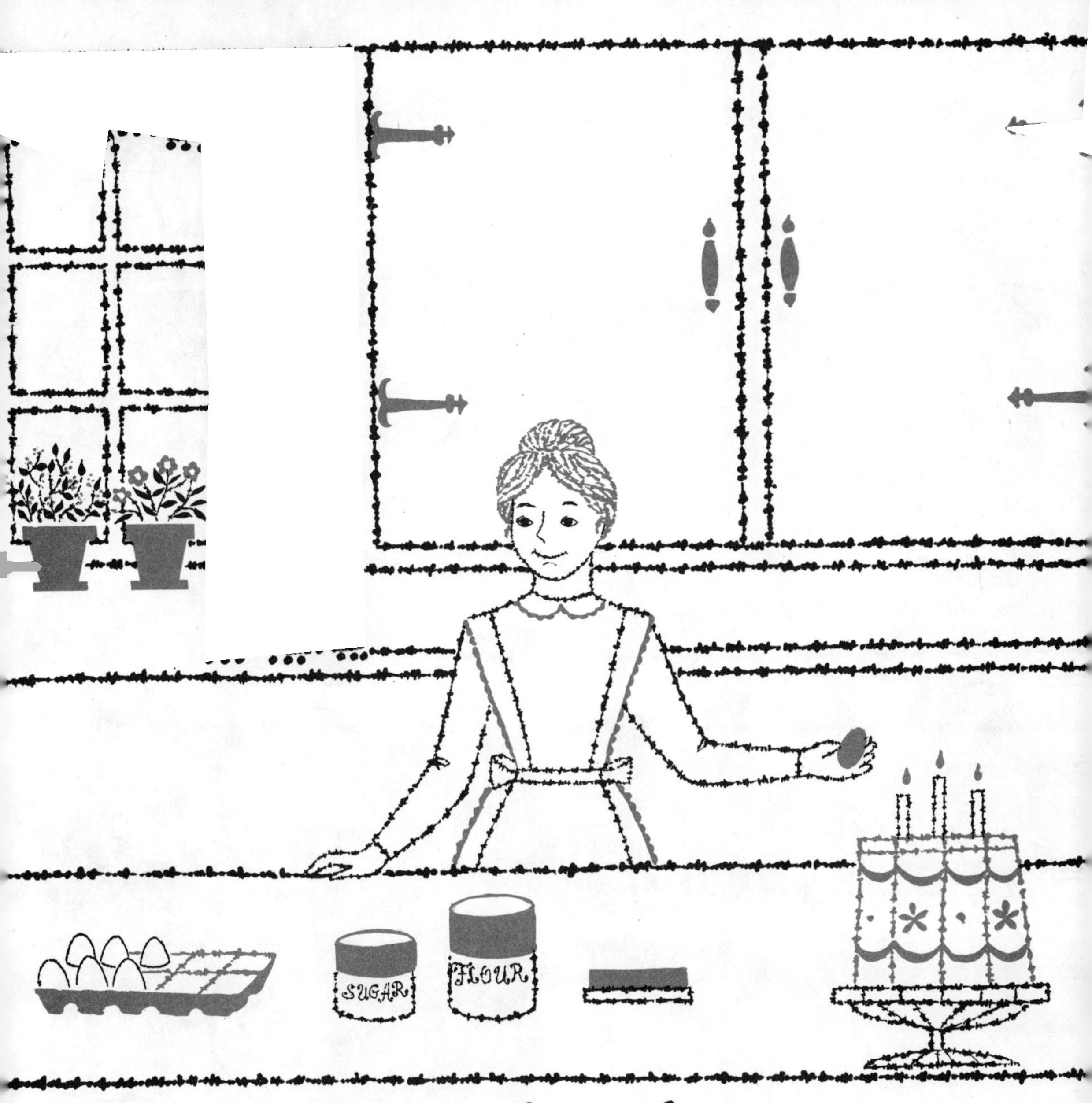

$10 - 4 = 6$

If she started with 10 eggs, then she subtracts 4 eggs. She has 6 eggs left.

A teacher uses subtraction to figure out how many more girls than boys there are in a class. If there are 8 boys in the class and 10 girls, the teacher subtracts 8 from 10. Then she knows she has 2 more girls than boys in the class.

And you use subtraction to figure out how much money you will have left if you spend 10 cents for candy.

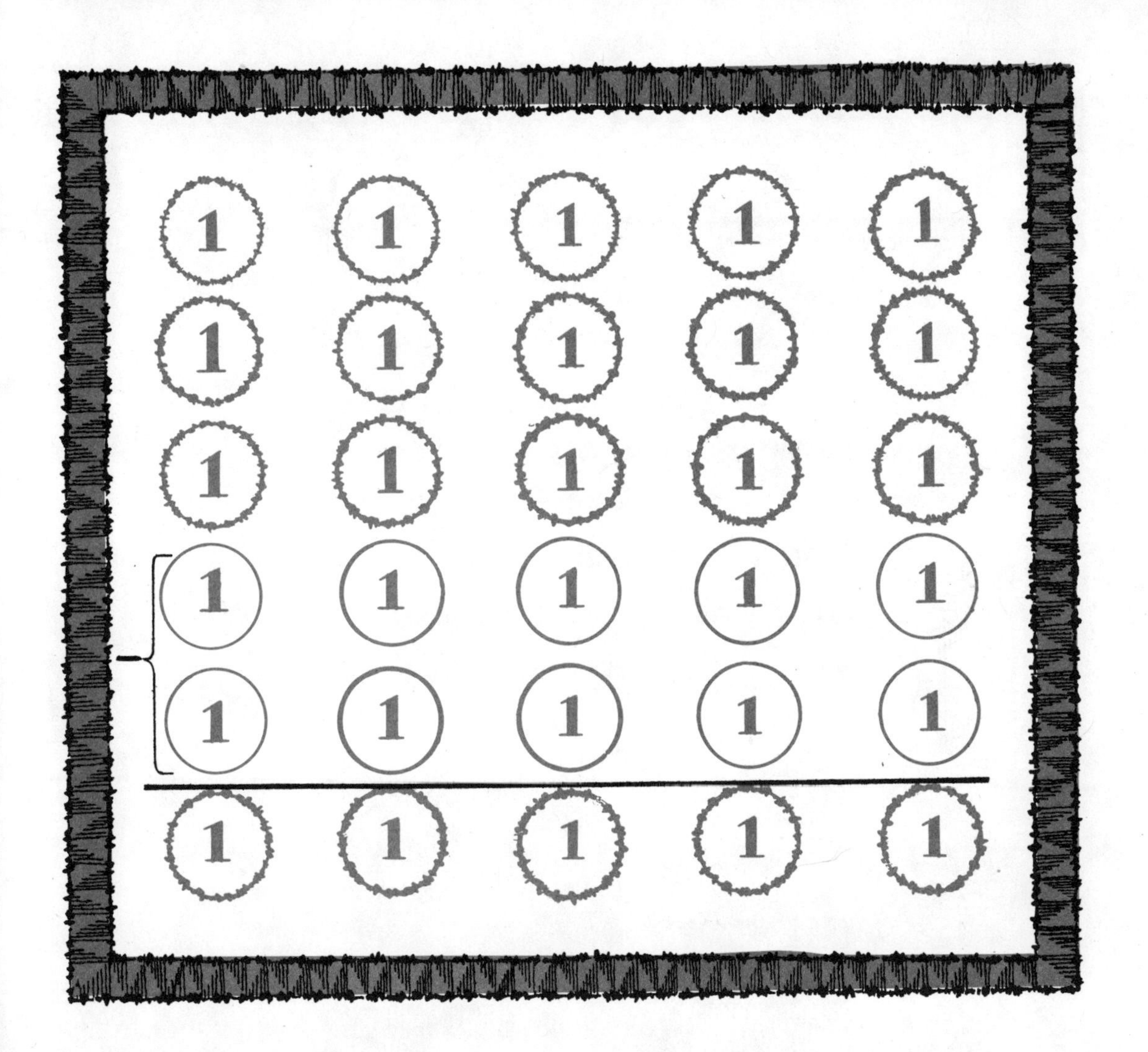

If you had 15 cents you subtract the 10 cents
you spent from the 15 cents you had.
You have 5 cents left.

1c 1c 1c 1c

One way to subtract is to count.
Suppose you have 7 cents.
How many cents will you
have left if you take away 5 cents?
You can count five into one pile
and then count how many are left.
You have 2 cents left.

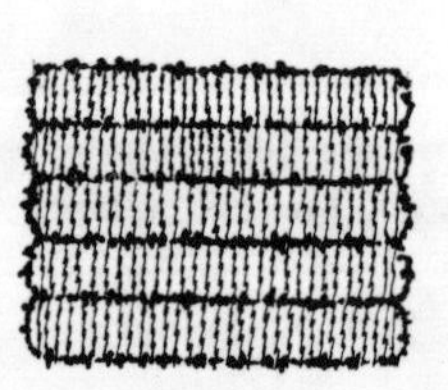

$$7 - 5 = 2$$

1c 1c 1c

You have 6 pencils.

If you subtract 2 pencils,
how many will be left?

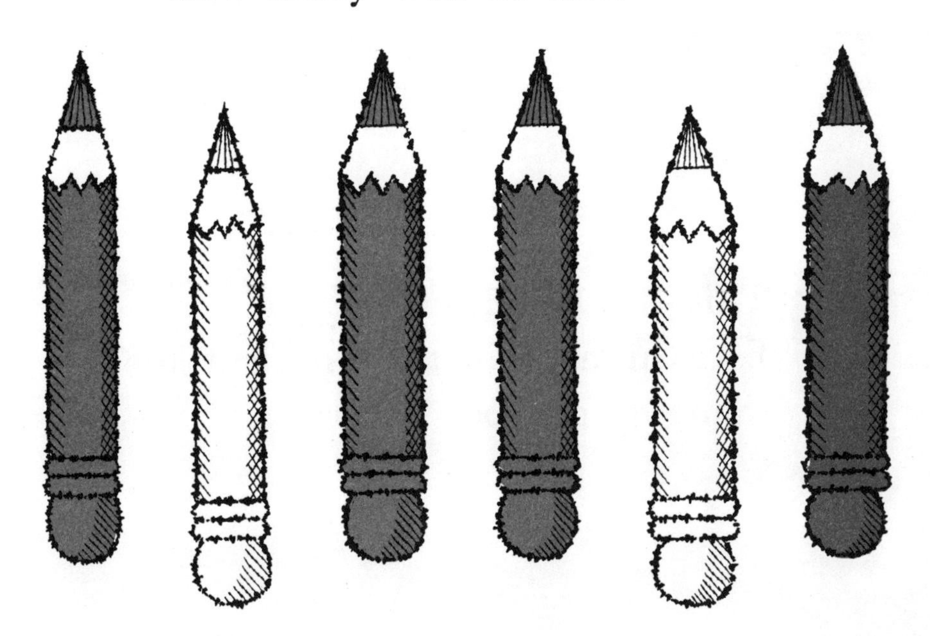

$6 - 2 = 4$

Another way to subtract is to use a number line.

1 2 3 4 5 6 7 8 9 10

This is a number line.
Each number in the line is 1 more than the number before it.

Suppose you want to subtract 4 from 9. First, find 9 on the number line. Then, starting with 9, count 4 to the left, or to the smaller numbers. You find that you have 5 remaining.

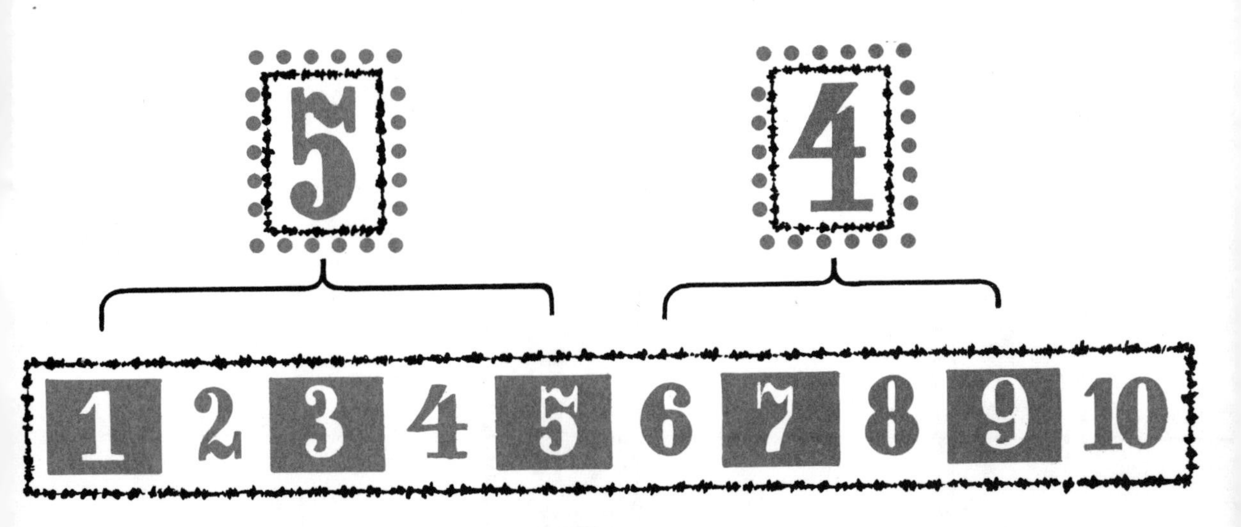

Then you know that

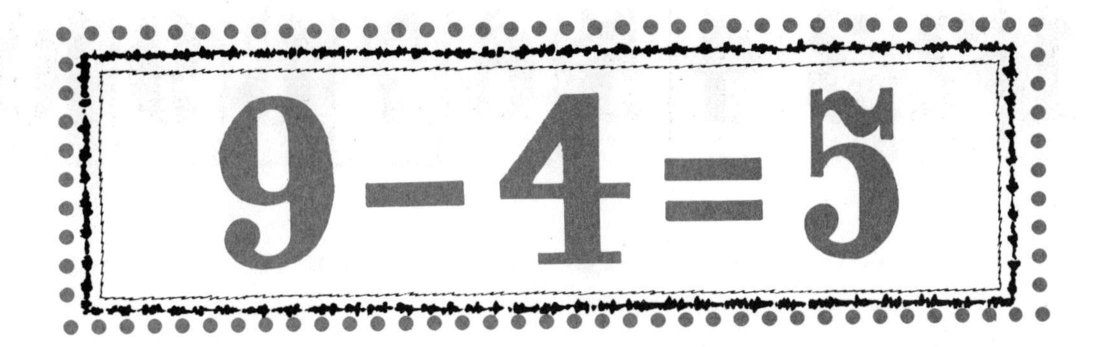

Subtract 5 from 8.
First, find 8 on the number line.
Next, starting with 8, count 5 to the left.
You have 3 remaining.
So you know that $8 - 5 = 3$.

You can use a longer number line to subtract bigger numbers.

Subtract 7 from 13.
First, find 13 on the number line.
Next, starting with 13, count 7 to the left.

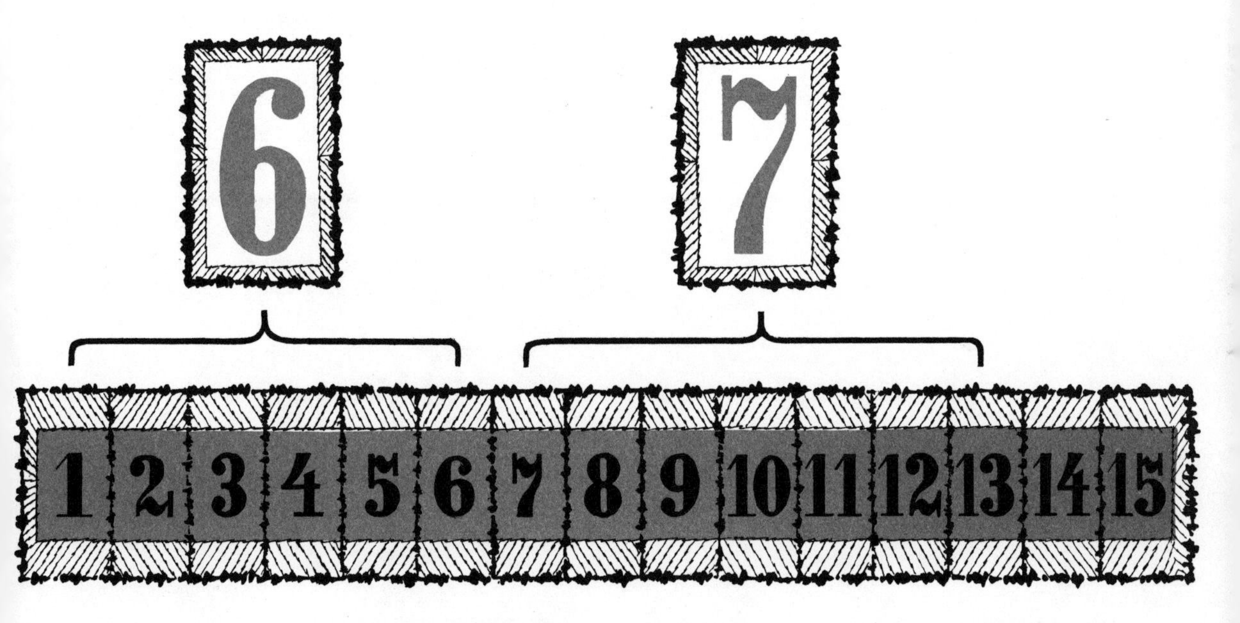

You have 6 remaining.
So you know that $13-7=6$.

How much is

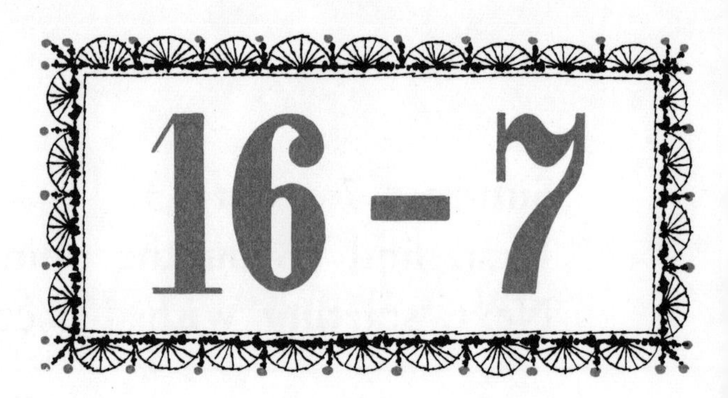

If you use the number line, you can see that

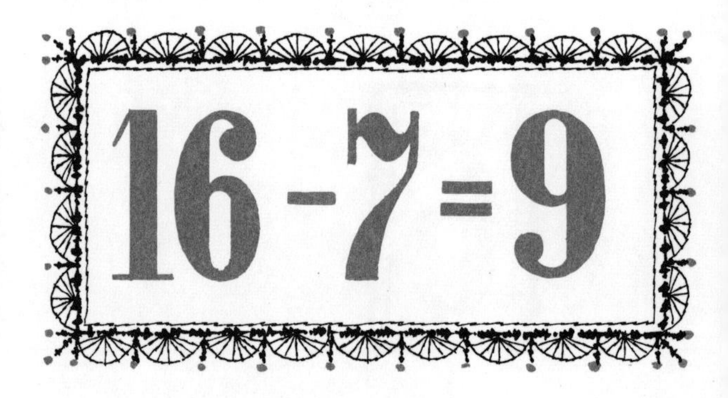

But it is much quicker to subtract by

than by counting or by using a number line.

You already know how to use addition. So you know that

$$3 + 4 = 7$$

or

$$2 + 3 = 5$$

or

$$9 + 6 = 15$$

When you know addition, it can help you to do subtraction by thinking.

If you have the problem

you can think of the addition fact

Then you know that 3 is the answer to the subtraction problem.

If you have the problem

you can think of the addition fact

Then you know that 2 is the answer to the subtraction problem.

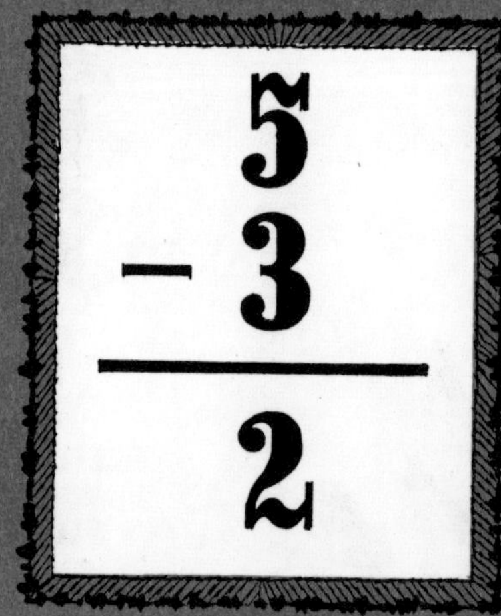

If you have

$$\begin{array}{r} 15 \\ -\ 9 \end{array}$$

Think of

$$\begin{array}{r} 9 \\ +\ 6 \\ \hline 15 \end{array}$$

Then you know that 6 is the answer to the subtraction problem.

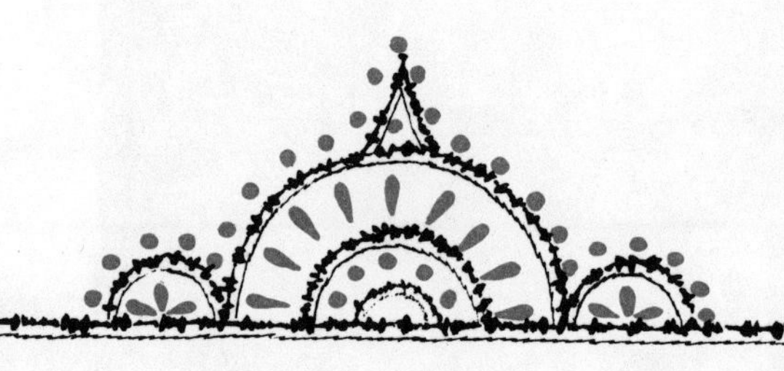

We call these the subtraction facts.

You can try each of them on a number line.

If you learn them, you can do subtraction just by thinking.

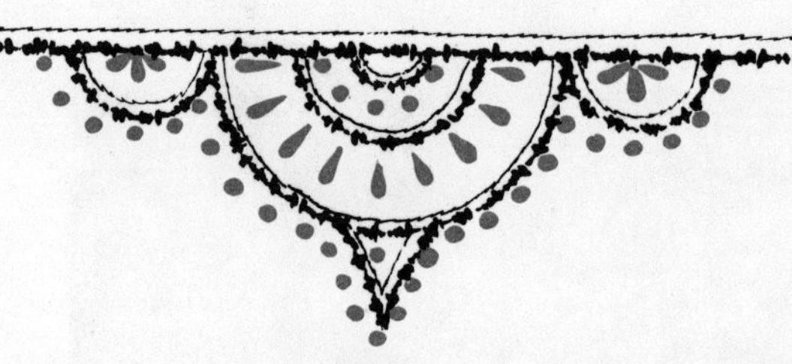

$$\begin{array}{rrrrrrrrrr} 1 & 2 & 3 & 4 & 5 & 6 & 7 & 8 & 9 & 10 \\ -1 & -1 & -1 & -1 & -1 & -1 & -1 & -1 & -1 & -1 \\ \hline 0 & 1 & 2 & 3 & 4 & 5 & 6 & 7 & 8 & 9 \end{array}$$

$$\begin{array}{rrrrrrrrrr} 2 & 3 & 4 & 5 & 6 & 7 & 8 & 9 & 10 & 11 \\ -2 & -2 & -2 & -2 & -2 & -2 & -2 & -2 & -2 & -2 \\ \hline 0 & 1 & 2 & 3 & 4 & 5 & 6 & 7 & 8 & 9 \end{array}$$

$$\begin{array}{rrrrrrrrrr} 3 & 4 & 5 & 6 & 7 & 8 & 9 & 10 & 11 & 12 \\ -3 & -3 & -3 & -3 & -3 & -3 & -3 & -3 & -3 & -3 \\ \hline 0 & 1 & 2 & 3 & 4 & 5 & 6 & 7 & 8 & 9 \end{array}$$

$$\begin{array}{rrrrrrrrrr} 4 & 5 & 6 & 7 & 8 & 9 & 10 & 11 & 12 & 13 \\ -4 & -4 & -4 & -4 & -4 & -4 & -4 & -4 & -4 & -4 \\ \hline 0 & 1 & 2 & 3 & 4 & 5 & 6 & 7 & 8 & 9 \end{array}$$

$$\begin{array}{rrrrrrrrrr} 5 & 6 & 7 & 8 & 9 & 10 & 11 & 12 & 13 & 14 \\ -5 & -5 & -5 & -5 & -5 & -5 & -5 & -5 & -5 & -5 \\ \hline 0 & 1 & 2 & 3 & 4 & 5 & 6 & 7 & 8 & 9 \end{array}$$

$$\begin{array}{rrrrrrrrrr} 6 & 7 & 8 & 9 & 10 & 11 & 12 & 13 & 14 & 15 \\ -6 & -6 & -6 & -6 & -6 & -6 & -6 & -6 & -6 & -6 \\ \hline 0 & 1 & 2 & 3 & 4 & 5 & 6 & 7 & 8 & 9 \end{array}$$

$$\begin{array}{rrrrrrrrrr} 7 & 8 & 9 & 10 & 11 & 12 & 13 & 14 & 15 & 16 \\ -7 & -7 & -7 & -7 & -7 & -7 & -7 & -7 & -7 & -7 \\ \hline 0 & 1 & 2 & 3 & 4 & 5 & 6 & 7 & 8 & 9 \end{array}$$

$$\begin{array}{rrrrrrrrrr} 8 & 9 & 10 & 11 & 12 & 13 & 14 & 15 & 16 & 17 \\ -8 & -8 & -8 & -8 & -8 & -8 & -8 & -8 & -8 & -8 \\ \hline 0 & 1 & 2 & 3 & 4 & 5 & 6 & 7 & 8 & 9 \end{array}$$

$$\begin{array}{rrrrrrrrrr} 9 & 10 & 11 & 12 & 13 & 14 & 15 & 16 & 17 & 18 \\ -9 & -9 & -9 & -9 & -9 & -9 & -9 & -9 & -9 & -9 \\ \hline 0 & 1 & 2 & 3 & 4 & 5 & 6 & 7 & 8 & 9 \end{array}$$

To help learn the subtraction facts
you should copy the first two
numerals of each fact on a
separate sheet of paper or card.
Do NOT copy the answer on the sheet
of paper or card.

Next, ask someone in your family or a friend to hold up each paper or card, one at a time, while you say the whole subtraction fact.
Keep practicing until you can get each fact right every time.

When you have learned
all the subtraction facts,
you will be able to do harder
subtraction problems quickly
and easily.

Before you work subtraction problems, there are some new words you should learn.

The large number that you subtract from is called the

MINUEND

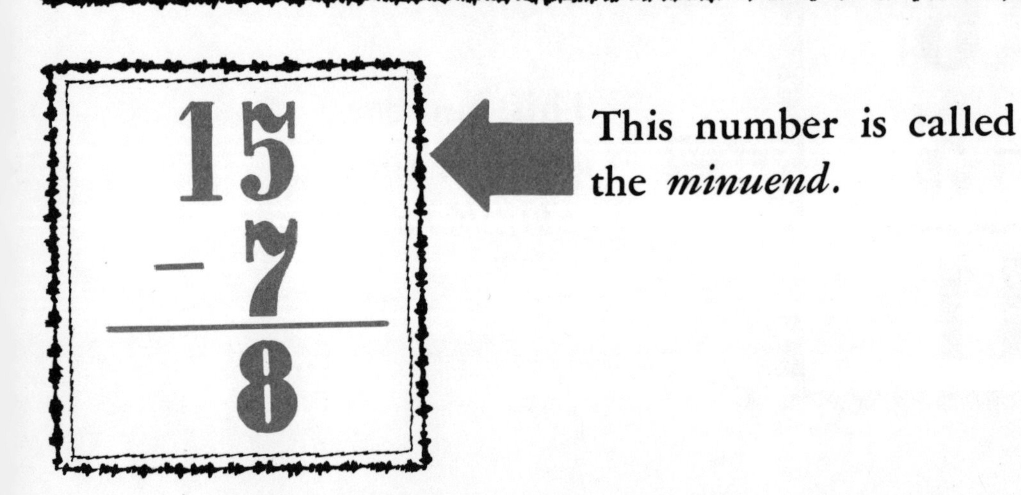

This number is called the *minuend*.

You say *minuend* as though it were spelled MIN-you-end.

The smaller number that you subtract from the minuend is called the

SUBTRAHEND

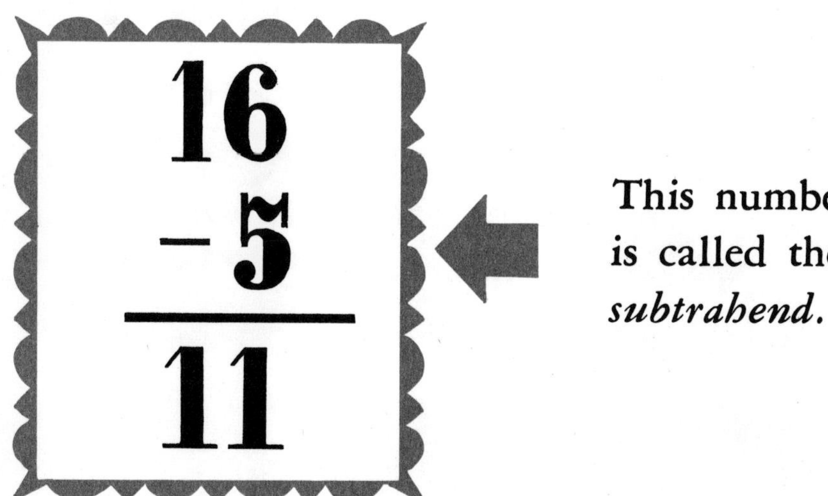

This number is called the *subtrahend.*

You say *subtrahend* as though it were spelled SUB-truh-hend.

When you subtract, or take away, a number from another number, the answer is called the

REMAINDER

Sometimes it is also called the

DIFFERENCE

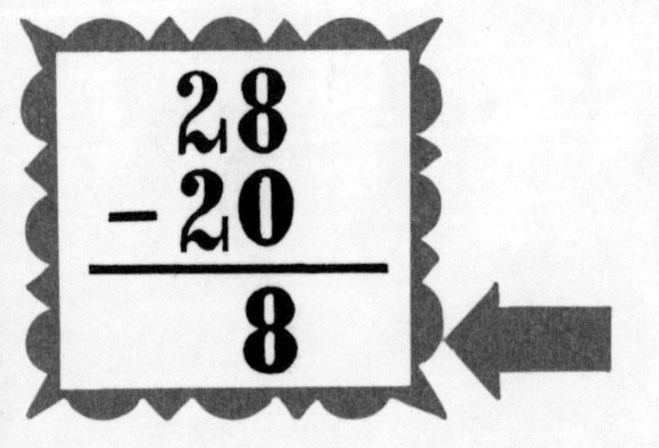

This number is called the *remainder*, or the *difference*.

You say *remainder* as though it were spelled rih-MAYN-duhr. You say *difference* as though it were spelled DIF-uhr-ents.

If you have learned the subtraction facts, it is not hard to subtract large numbers.

Suppose you have the problem

$$\begin{array}{r} 98 \\ -34 \\ \hline \end{array}$$

First, think of 98 as 9 tens and 8 ones.

Next, think of 34 as 3 tens and 4 ones.

Now subtract the ones.

$$\begin{array}{r} 8 \text{ ones} \\ -4 \text{ ones} \\ \hline 4 \text{ ones} \end{array}$$

You know the answer is 4 from your subtraction facts.

Then subtract the tens.

$$\begin{array}{r} 9 \text{ tens} \\ -3 \text{ tens} \\ \hline 6 \text{ tens} \end{array}$$

You know the answer is 6 from your subtraction facts.

Let's put the problem back together again to see the whole answer. The answer is 6 tens and and 4 ones, or 64,

9 tens and 8 ones		98
	or	
−3 tens and 4 ones		−34
6 tens and 4 ones		64

Here are four subtraction problems with large numbers. Can you work them?
Copy the problems on a piece of paper or card.
Remember to separate the numbers into tens and ones.

AND ALWAYS SUBTRACT THE ONES FIRST.

1. $$\begin{array}{r} 74 \\ -32 \\ \hline \end{array} \qquad \begin{array}{r} 7 \text{ tens and } 4 \text{ ones} \\ -3 \text{ tens and } 2 \text{ ones} \\ \hline \end{array}$$

2. $$\begin{array}{r} 37 \\ -15 \\ \hline \end{array} \qquad \begin{array}{r} 3 \text{ tens and } 7 \text{ ones} \\ -1 \text{ ten and } 5 \text{ ones} \\ \hline \end{array}$$

3. $$\begin{array}{r} 95 \\ -64 \\ \hline \end{array} \qquad \begin{array}{r} 9 \text{ tens and } 5 \text{ ones} \\ -6 \text{ tens and } 4 \text{ ones} \\ \hline \end{array}$$

4. $$\begin{array}{r} 89 \\ -43 \\ \hline \end{array} \qquad \begin{array}{r} 8 \text{ tens and } 9 \text{ ones} \\ -4 \text{ tens and } 3 \text{ ones} \\ \hline \end{array}$$

After you have worked the problems, check your answers with these correct answers:

1.	74	7 tens and 4 ones
	−32	−3 tens and 2 ones
	42	4 tens and 2 ones
2.	37	3 tens and 7 ones
	−15	−1 ten and 5 ones
	22	2 tens and 2 ones
3.	95	9 tens and 5 ones
	−64	−6 tens and 4 ones
	31	3 tens and 1 one
4.	89	8 tens and 9 ones
	−43	−4 tens and 3 ones
	46	4 tens and 6 ones

It is not hard to subtract larger numbers. Separate them into hundreds, tens, and ones.

594		5 hundreds	9 tens	4 ones
	or			
−283		−2 hundreds	−8 tens	−3 ones

First, subtract the ones:

5 hundreds	9 tens	4 ones
−2 hundreds	−8 tens	−3 ones
		1 one

Next, subtract the tens:

5 hundreds	9 tens	4 ones
−2 hundreds	−8 tens	−3 ones
	1 ten	1 one

Finally, subtract the hundreds:

5 hundreds	9 tens	4 ones	
−2 hundreds	−8 tens	−3 ones	
3 hundreds	1 ten	1 one	or

311

ALWAYS REMEMBER TO SUBTRACT THE ONES FIRST, NEXT THE TENS, AND FINALLY THE HUNDREDS.

Here are four subtraction problems in the hundreds.

Can you work them?

Copy the problems on a piece of paper or a card.

1.	974 or -621		9 hundreds -6 hundreds	7 tens -2 tens	4 ones -1 one
2.	649 or -418		6 hundreds -4 hundreds	4 tens -1 ten	9 ones -8 ones
3.	868 or -534		8 hundreds -5 hundreds	6 tens -3 tens	8 ones -4 ones
4.	793 or -281		7 hundreds -2 hundreds	9 tens -8 tens	3 ones -1 one

After you have worked the problems, check your answers with these correct answers:

1.	974	or	9 hundreds	7 tens	4 ones
	−621		−6 hundreds	−2 tens	−1 one
	353		3 hundreds	5 tens	3 ones
2.	649	or	6 hundreds	4 tens	9 ones
	−418		−4 hundreds	−1 ten	−8 ones
	231		2 hundreds	3 tens	1 one
3.	868	or	8 hundreds	6 tens	8 ones
	−534		5 hundreds	−3 tens	−4 ones
	334		3 hundreds	3 tens	4 ones
4.	793	or	7 hundreds	9 tens	3 ones
	−281		−2 hundreds	−8 tens	−1 one
	512		5 hundreds	1 ten	2 ones

Now suppose you have the subtraction problem

$$\begin{array}{r} 52 \\ -24 \\ \hline \end{array}$$

When you separate this problem into tens and ones,

$$\begin{array}{rr} 5 \text{ tens} & 2 \text{ ones} \\ -2 \text{ tens} & 4 \text{ ones} \\ \hline \end{array}$$

you cannot subtract because there are not enough ones in the minuend. (Remember that the minuend is the number you subtract from. It must be larger than the subtrahend.)

To work this problem you must write the numbers in the minuend in a different way. You can do this when you know that

1 ten is the same as 10 ones

or

1 ten is made up of 10 ones.

52 is 5 tens and 2 ones. If you take 1 of the tens away you have 4 tens. Now you change the 1 ten into 10 ones. You add the 10 ones to the 2 ones. Now you have 4 tens and 12 ones.

52	or	5 tens	2 ones	or	4 tens	12 ones
−24		−2 tens	−4 ones		−2 tens	−4 ones

Now it is easy to work the problem. You can find the answer is 2 tens and 8 ones or 28.

4 tens	12 ones		52
−2 tens	−4 ones	or	−24
−2 tens	8 ones		28

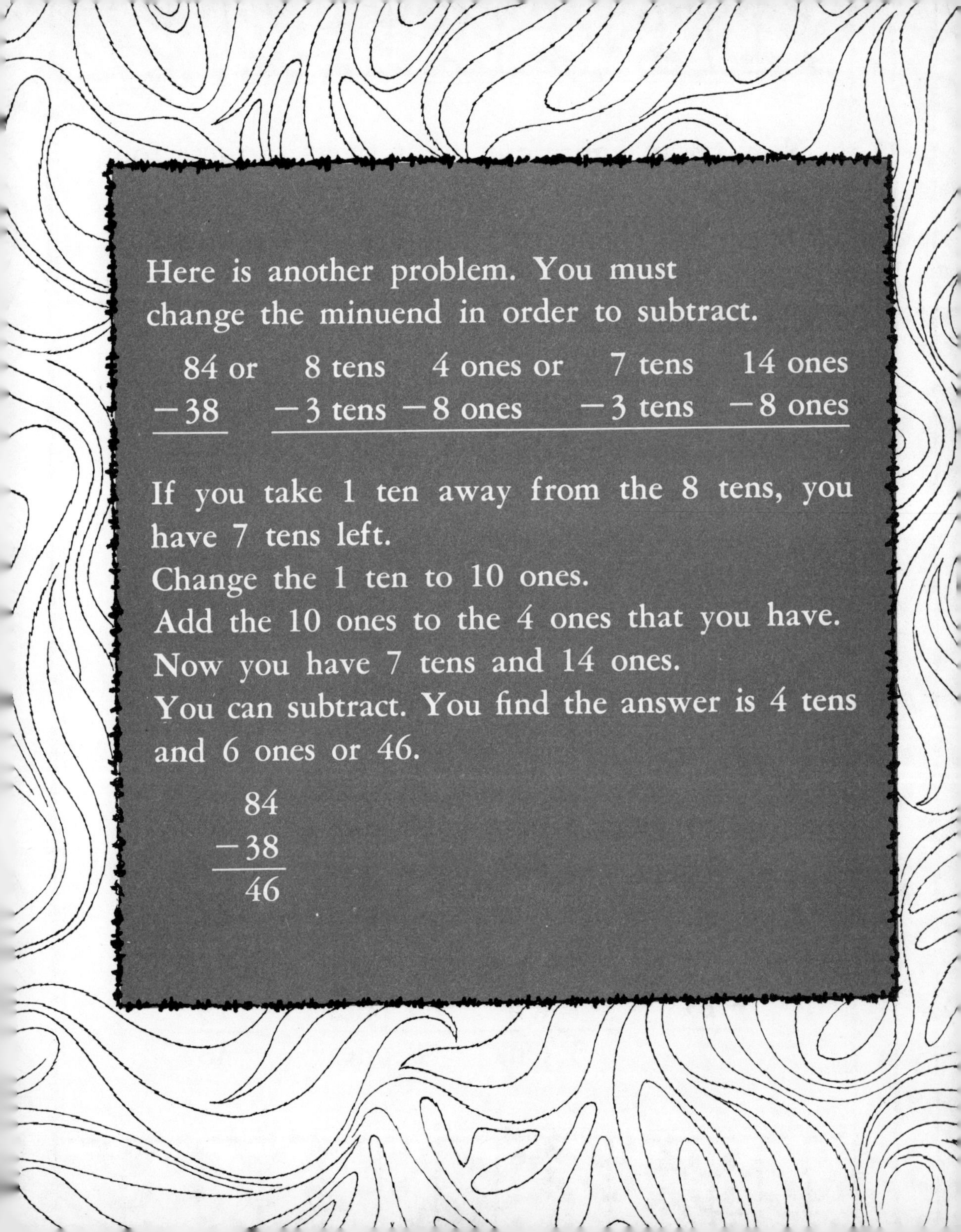

Here is another problem. You must change the minuend in order to subtract.

$$\begin{array}{r} 84 \\ -38 \\ \hline \end{array} \text{ or } \begin{array}{rr} 8 \text{ tens} & 4 \text{ ones} \\ -3 \text{ tens} & -8 \text{ ones} \\ \hline \end{array} \text{ or } \begin{array}{rr} 7 \text{ tens} & 14 \text{ ones} \\ -3 \text{ tens} & -8 \text{ ones} \\ \hline \end{array}$$

If you take 1 ten away from the 8 tens, you have 7 tens left.
Change the 1 ten to 10 ones.
Add the 10 ones to the 4 ones that you have.
Now you have 7 tens and 14 ones.
You can subtract. You find the answer is 4 tens and 6 ones or 46.

$$\begin{array}{r} 84 \\ -38 \\ \hline 46 \end{array}$$

Now try to figure out the answers to these four problems.

Be sure to change the numbers in the minuend. Check your answers with the correct answers at the bottom of this page.

1.	2.	3.	4.
81	74	65	53
−37	−26	−28	−15

Here are the correct answers:

1.	81	or	7 tens	11 ones	or	81
	−37		−3 tens	−7 ones		−37
			4 tens	4 ones		44
2.	74	or	6 tens	14 ones		74
	−26		−2 tens	−6 ones		−26
			4 tens	8 ones		48
3.	65	or	5 tens	15 ones		65
	−28		−2 tens	−8 ones		−28
			3 tens	7 ones		37
4.	53	or	4 tens	13 ones		53
	−15		−1 ten	−5 ones		−15
			3 tens	8 ones		38

Now suppose you have the subtraction problem

$$\begin{array}{r} 435 \\ -286 \\ \hline \end{array}$$

When you separate this problem into hundreds, tens, and ones

4 hundreds 3 tens 5 ones
2 hundreds 8 tens 6 ones

you cannot subtract because there are not enough tens and ones in the minuend.

To work this problem you must write the numbers in the minuend in a different way. You know that

1 ten is made up of 10 ones

You must also know that

1 hundred is made up of 10 tens

435 is 4 hundreds 3 tens and 5 ones. First change 1 ten into 10 ones and add it to the 5 ones. Now you have

4 hundreds 2 tens and 15 ones

Then change 1 hundred into 10 tens and add it to the 2 tens. Now you have

3 hundreds 12 tens and 15 ones

Now it is easy to work the problem. You can find the answer is

3 hundreds	12 tens	15 ones		435
−2 hundreds	−8 tens	−6 ones	or	−286
1 hundred	4 tens	9 ones		149

Here are four problems. You mu
ones, tens, and hundreds in the m
you can subtract. Check your ans
correct answers at the bottom of

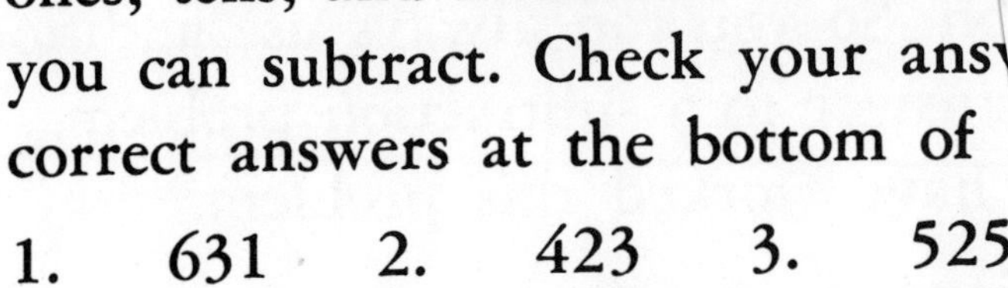

1.	631	2.	423	3.	525		
	−445		−247		−348		−268

Here are the correct answers:

1.	631 or	5 hundreds	12 tens	11 ones
	−445	4 hundreds	4 tens	5 ones
	186	1 hundred	8 tens	6 ones
2.	423 or	3 hundreds	11 tens	13 ones
	−247	2 hundreds	4 tens	7 ones
	176	1 hundred	7 tens	6 ones
3.	525 or	4 hundreds	11 tens	15 ones
	−348	3 hundreds	4 tens	8 ones
	177	1 hundred	7 tens	7 ones
4.	736 or	6 hundreds	12 tens	16 ones
	−268	2 hundreds	6 tens	8 ones
	468	4 hundreds	6 tens	8 ones

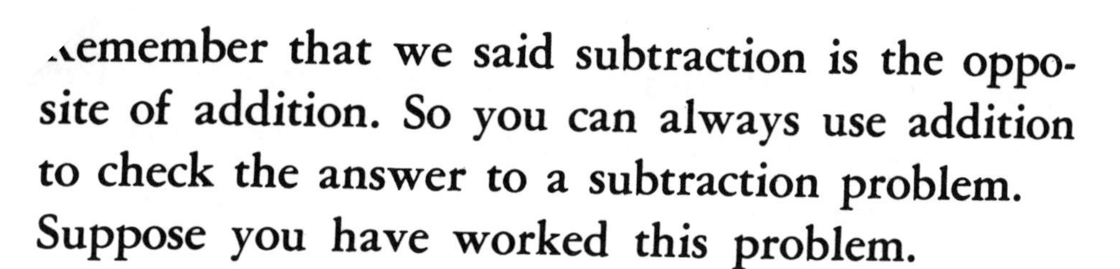

Remember that we said subtraction is the opposite of addition. So you can always use addition to check the answer to a subtraction problem. Suppose you have worked this problem.

932	minuend
−811	subtrahend
121	

You can check your answer if you add the subtrahend and the remainder.

$$\begin{array}{r} 811 \\ +121 \\ \hline \end{array}$$

The answer that you get should be the same as the minuend. Then you know that your answer to the subtraction problem is correct.

minuend	932	811	subtrahend
subtrahend	−811	+121	remainder
remainder	121	932	answer

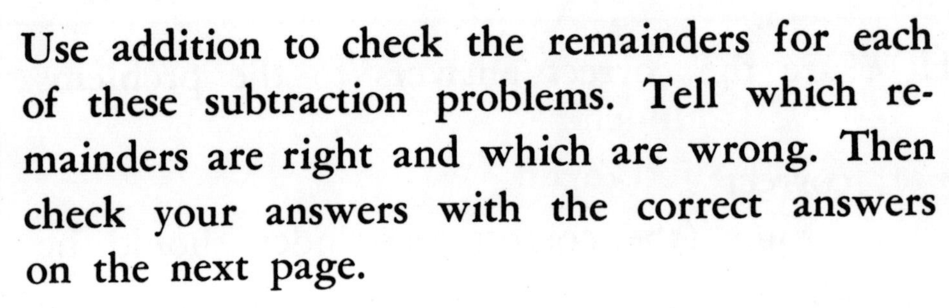

Use addition to check the remainders for each of these subtraction problems. Tell which remainders are right and which are wrong. Then check your answers with the correct answers on the next page.

1. $\begin{array}{r} 56 \\ -24 \\ \hline 32 \end{array}$ 2. $\begin{array}{r} 354 \\ -163 \\ \hline 291 \end{array}$ 3. $\begin{array}{r} 89 \\ -41 \\ \hline 48 \end{array}$

4. $\begin{array}{r} 93 \\ -27 \\ \hline 66 \end{array}$ 5. $\begin{array}{r} 76 \\ -38 \\ \hline 48 \end{array}$ 6. $\begin{array}{r} 289 \\ -94 \\ \hline 195 \end{array}$

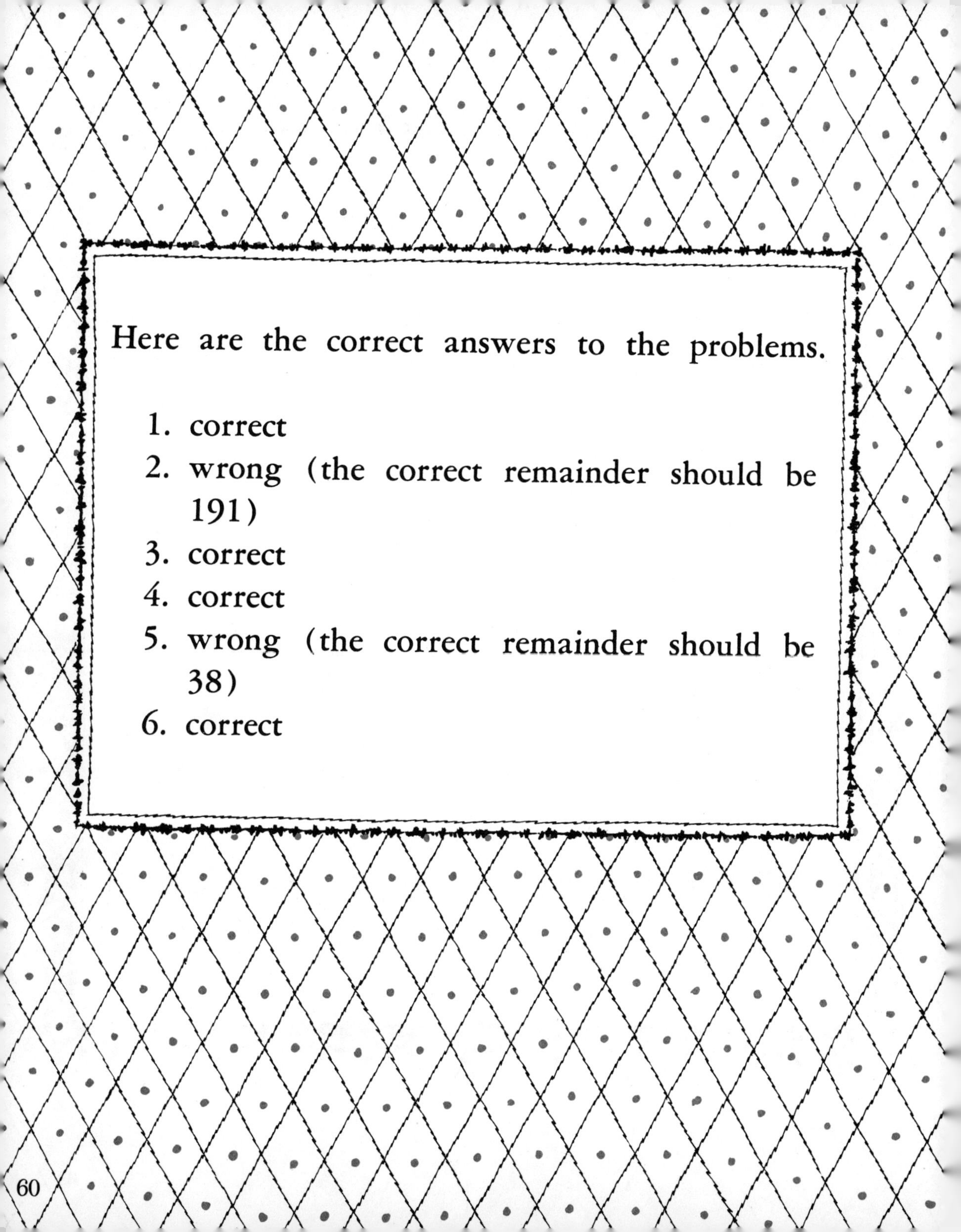

Here are the correct answers to the problems.

1. correct
2. wrong (the correct remainder should be 191)
3. correct
4. correct
5. wrong (the correct remainder should be 38)
6. correct

Now that you can work subtraction problems with numbers, you can do subtraction word problems, too.

If you had 75 cents and you spent 15 cents to buy an ice cream cone, how much money would you have left?

Think to yourself:

75 cents is the money I had.

15 cents is how much money I spent.

So I must subtract 15 cents from 75 cents to find out how much money I would have left.

$$\begin{array}{r} 75 \text{ cents} \\ -\,15 \text{ cents} \\ \hline 60 \text{ cents} \end{array}$$

You find that 60 cents is the remainder, so 60 cents is the amount of money you would have left.

Here are some more subtraction word problems for you to work. Check your answers against the correct answers on the next page.

1. Bob had 138 toy soldiers. Bill had 64 toy soldiers. How many more toy soldiers did Bob have than Bill?

2. A grocery store had 45 baskets of apples. In one day the store sold 28 baskets of apples. How many baskets of apples were left?

3. Mr. Jones' car went 237 miles in one day. Mr. Black's car went 164 miles in one day. How much farther did Mr. Jones' car go than Mr. Black's car?

4. Jill had 95 cents. She bought a magazine for 75 cents. How much did she have left?

Here are the correct answers for the subtracti
word problems on the last page.

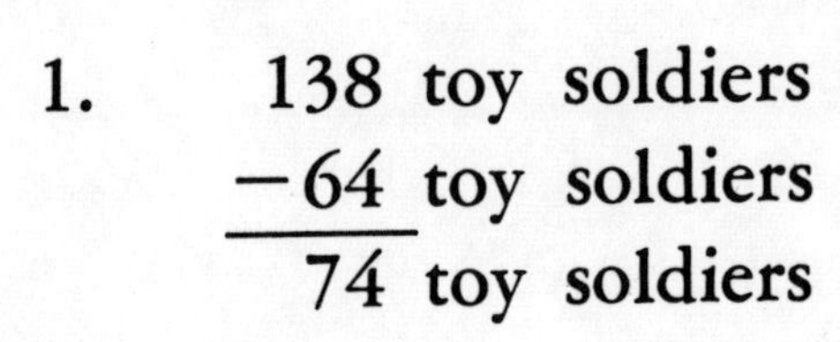

1. 138 toy soldiers
−64 toy soldiers
74 toy soldiers

2. 45 baskets of apples
−28 baskets of apples
17 baskets of apples

3. 237 miles
−164 miles
73 miles

4. 95 cents
−75 cents
20 cents